People Travel in All Kinds of Weather

by Susan Halko

People travel in all kinds of weather.

50
305

People travel in **rainy** weather.

Some people travel by truck.

Some people travel by train.

People travel in **sunny** weather.

Some people travel by car.

Some people travel by bike.

People travel in **windy** weather.

Some people travel by boat.

Some people travel by glider.

People travel in **cloudy** weather.

Some people travel by jet.

Some people travel by bus.

People Travel in All Kinds of Weather

rainy

rainy

sunny

sunny

windy

windy

cloudy

cloudy

Angelina's Christmas

Illustrations by Helen Craig Story by Katharine Holabird

A TRUMPET CLUB SPECIAL EDITION

Christmas was coming, and everyone at Angelina's school worked hard to prepare for the Christmas show. Angelina and the other children stayed after their classes to rehearse and help decorate the hall.

When Angelina left school it was already dark outside. Large snowflakes were falling and Angelina was so excited that she danced along the pavement.

The cottages in the village looked warm and welcoming, with holly wreaths on the doors and Christmas lights shining in all the windows. But the very last cottage was cold and dark. Angelina peeped in the window and saw an old man huddled by a tiny fire.

Angelina ran the rest of the way home and found her mother and little cousin Henry in the kitchen. She asked her mother about the man who lived all alone in the cottage.

"Oh, that's Mr. Bell," her mother replied. "He used to be the postman, but he's too old to work now."

Angelina wanted to make a Christmas surprise for Mr. Bell, so Mrs. Mouseling gave her some dough to make cookies shaped like stars, bells and trees.

Henry had a piece of dough too, and he made a nice big Santa Claus cookie. “Look!” he said proudly. “I’m going to see Santa tonight and give him this cookie *myself!*”

“Santa only comes very late at night after everyone has gone to bed,” Angelina explained. “Why don’t you leave your cookie out on a plate for him?”

Henry started to cry. "No, no, no. I want to see Santa Claus!"

"Don't be such a crybaby, Henry," Angelina scolded, but Henry didn't stop crying.

The next afternoon Angelina and her mother packed a basket with the cookies and some mince pies and fruit for Mr. Bell. "Don't you want to help Angelina take the presents to Mr. Bell?" asked Mrs. Mouseling, but Henry only shook his head.

Then Angelina and her father went out to find a Christmas tree for Mr. Bell. Henry followed Angelina and Mr. Mouseling all the way to Mr. Bell's cottage. He still wouldn't say a word.

The old postman was amazed and delighted to see his visitors. He invited Angelina and her father inside, and then he noticed Henry standing alone in the snow. "Come in, my friend!" said Mr. Bell with a smile, and he picked Henry up and brought him in near the fire.

Mr. Bell's eyes were bright and twinkling. "Wait here a moment," he said, and disappeared up the stairs. Then he came down looking . . .

just like SANTA CLAUS!

"This is the red costume I wore once when Santa Claus needed someone to take his place at the village Christmas party," said Mr. Bell with a chuckle, and he sat down and took Henry on his knee. While Mr. Mouseling made tea and Angelina decorated the tree, Henry listened to Mr. Bell's stories.

"I used to go out on my bicycle, no matter what the weather, to deliver presents to all the children in the countryside. One year there was a terrible blizzard and all the roads were covered with snow. I had to deliver the toys on a sled, and if I hadn't glimpsed the village lights blinking in the distance I would have been lost out in the storm." Henry listened with wide eyes.

When it was time to go Henry reached into his pocket. "I made this," he said. "It's for you." Out of his pocket he took his big Santa Claus cookie and gave it to Mr. Bell.

"This is the best Christmas surprise I've had for many years," said Mr. Bell, thanking Henry and Angelina for their presents. Angelina said she wished Mr. Bell would come to her school show in his Santa Claus costume.

"That would be a pleasure," he said, smiling.

Mr. Bell kept his promise: he came to the Christmas show in his red costume and watched Angelina and

her friends dressed as sugar plum fairies dancing the Nutcracker Suite.

Later all the children gathered around Mr. Bell, and Henry felt proud as Mr. Bell handed out the Christmas presents and entertained everyone with stories about his adventures as a postman.

Mr. Bell was never lonely at Christmas again, because every year he was invited to come to Angelina's school in his Santa Claus suit for the Christmas show.